ORANGE JUICE
EXTRACTOR MACHINES

Benedict Nnamdi Ugwu

ELIVA PRESS

ELIVA PRESS

Benedict Nnamdi Ugwu

Orange juice contains many nutrients required by individuals such as vitamins A, B, C and so on. Therefore, consuming oranges through fresh-made juice allows you to circumvent the digestion process and deliver concentrated nutrients into your bloodstream. It is also easier to consume larger amounts of oranges using juice extractor machine to meet our daily requirements. A nutritionist once said that the key benefit of using a juice extractor at home is that you are in control of the ingredients that go into your juice. Juice extraction has been a worrisome problem to local farmers in Nigeria due to their perishable nature of their produce. The inability of local farmers to afford the high cost of imported juice extractor has worsened the problem. Hence, both mechanized and manual fruit juice machines discussed in this work is developed to provide an affordable and user friendly machine. These machine models exist all over the world with very few becoming popular while the rest got fizzle out due to their limitations. There is a great need to analyze the orange juice extractor machines that squeeze the juices from the orange for better and healthy juice extraction.

Published: Eliva Press SRL
Address: MD-2060, bd.Cuza-Voda, 1/4, of. 21 Chişinău, Republica Moldova
Email: info@elivapress.com
Website: www.elivapress.com

ISBN: 978-1-63648-028-2

ORANGE JUICE EXTRACTOR MACHINES

ACKNOWLEDGEMENT

We wish to acknowledge the management of Enugu State University of Science and Technology (ESUT) Enugu, Nigeria. More especially the staff and students of Mechanical and Production Engineering for their support and resilience. To my wife and children that remained steadyfast in their prayers I say more of God's grace.

Table of content

List of Figures (Illustrations)

ABSTRACT

There are some agricultural equipment for post harvesting of Orange fruit Juice, all the machine are geared towards extraction of the fruit juice. Fruit juice extraction is the act of wringing out the juice content of fruits by way of an effective processing and storage which enhance reduction in wastage. Fruit juices which literally have high antioxidants help in increasing serum capacity of the body and at the same time balances the oxidative stress and discomfort normally caused by high-fatty and sugar meals. The history of juice extraction dates back to the nineteenth century. The extraction of juice from its fruit has progressed tremendously from the old tedious method of squeezing to an automated juice extracting machine across the world, making it an essential tool for citrus farmers. Juice extractor machine are classified broadly into four types centrifugal, masticating, Triturating and Press juicers which may be operated manually or electrically. Many attempts have been made to extract oranges using both manual and mechanized means. Traditionally the task of extraction of orange is easy but time consuming. The manual still requires the operator to remove the pulps and the seeds while the mechanized is a whole automated process made easy.

Both mechanized and manual fruit juice machines discussed in this work is developed to provide an affordable and user friendly machine. These machine models exist all over the world with very few becoming popular while the rest got fizzle out due to their limitations. Then simple portable machines that squeeze the juices from the orange are to be reviewed in this work for better and healthy juice extraction.

CHAPTER ONE

1. INTRODUCTION

A juicer is a machine that has the capacity of producing juice from fruits, leafy green and vegetables. Several kinds of fruits or vegetables extracted are dependent on the type of fruit juicer developed [1]. A juice extractor is an implement that can extract juice from both fruits and vegetables. A juice extractor is a machine designed and fabricated to snap out fluid (juice) from the fruit, either by squeezing, pressing or crushing for the purpose of drinking. The juice extractors are classified into different types based on their mode of operation.

Fruit juice extractor is an enhanced agricultural device which uses the pressing mechanism to extract juices from some fruit [2]. The operational units of a complete fruit juice extraction processes are: sorting, grading, rinsing, peeling, cutting, juice formulation, clarification, storage and packaging [2].

This practice of squeezing, pressing and crushing of fruits just to obtain the juice and reduce the draw-back of waste and pulp is referred to as fruit juice extraction. The orange fruit can be eaten raw, or possibly extract its juice or fragrant peel as produce. Approximately 70% of citrus productions in 2012 were as a result of sweet oranges. In 2014, countries like California and Florida in United States of America with Brazil have ubiquitous rate of production of oranges with 70.9 million metric tons of oranges grown worldwide. Orange extraction started with hand extraction of juice which is rather slow, tedious and unhygienic, the use of machine came into being as the demand for juice consumption increased [3]. The fruit is abundant in the production season and always very scarce and expensive during the off season. Attempts to store fruit in its fresh and natural form have failed due to lack of effective storage and preservation methods.

Processing this fruit into juice is a better way of storage, preservation and value addition. The benefits of using machine for extraction are: saves time, improves efficiency, increase capacity and reduced spoilage and waste [4]. Physical operated juice extractors have been developed for home use but it has limited output [5]. Generally, electric powered juice extractors have accessories like electric motor, switch, belt, pulley, gears and bearings and components parts like a hopper which introduces the fruit to the machine compressing chamber, a housing unit (compressing chamber) which incorporates an array of pressers exceptionally arranged, a strainer (for sieve off waste), a juice collector container and a waste bin built-in for the orange waste (pulps, skin and seeds) disposal.

As there are no better ways to preserving this orange fruits, preservation of the orange fruits through extraction of the juice has been established as the most effective technique so far. With this extraction mechanism fruit juices can be stored and preserved for months or even years before expiration [2].

A simple machine produced from locally sourced materials for extraction of juice from the orange fruits effectively and efficiently at an affordable price so as to encourage a healthy living through consumption of fruit juice and longer preservation of the fruit during its harvest was necessary.

CHAPTER TWO

2. JUICE EXTRACTOR AND ITS CLASSIFICATION

A juice extractor also known as a juicer, is an implement used for extracting juice from fruits, leafy greens and other types of vegetables in a practice referred to a juicing [6]. It crushes squeezes and grinds the juice out from its fleshy tissue [7].They are different types of juice extractors; each works best for some variety of fruits and vegetables. Presently, juice extractors are classified into four essential types: masticator, centrifuge, triturating juicers (twin gear), and press juicer [8].

2.1 Centrifugal Juicer

Centrifuges designed as the fastest and most affordable of all motorized fruit juice extractor has mesh chamber where sharp blades rotate at a super-sonic speed to slice the fruits and extract the juice from the pulp [9]. These machines have advanced features that can process juice within seconds. Normally with a large feeding chute it can chow down larger volumes of fruit items even without pre-processing, this juicer reduces time sent on preparation work. Due to their very high speed noise, heat, and oxidation of the juice are observed. The heat generated breaks down certain enzymes and nutrients, while plenty air introduced oxidizes the juice their by causing a loss of nutrient thus reducing the juice quality as well as the shelf life. The basic component parts are plunger, top cover, top latch assembly, top blade, bottom blade, basket, juice bowl and mesh screen Centrifugal juicers are produced by Breville, Omega, Hamilton, Black and Decker. They have speedy, short prepping, juicing, and cleaning time as advantage. They are simple to assemble/disassemble, compact for processing and also affordable.

2.2 Masticating Juicer (Cold Press Juicers)

A masticating juicer has a screw worm shaft that works to press and crush the orange into lesser bit before pressing it against the juice extraction compartment for juicing [9].Masticating juicers are also referred to as single auger juicers or slow juicer, since it takes long time to produce fruit juice and vegetable juice as weighed alongside centrifuges. These machines are in two main varieties namely vertical and horizontal masticating juicers. Vertically configured models have larger auger and feeding chute while the horizontal juicers have feeding chute with smaller footprints, but are prone to blockage due to the placement of the pulp ejector. To avoid these small chutes problems which emanates pre-cutting of every produce before feeding is paramount. Masticators serve a multifunctional purpose of a grocery processor and grinder. Mostly found in kitchens as juice maker and grain mills. Even with their slow speed, they perform very well on both hard and soft agricultural produce of kale and orange; with exceptional juice quality occasioned by heat absences which naturally destroy both the enzymes and antioxidants in the juice. Masticators are seen as the best juicers for leafy greens. Basic components of masticating juicers are pusher, hopper, auger, perforated screen, spinning brush and silicon brush. They have high juice yield with improved juice quality. They function with tough and yielding materials but are generally costly.

2.3 Triturating Juicer

The triturating juicer comprises of two gears coupled closer to each another, with the intention of crushing, grinding and extracting juice from the agricultural produce using a low speed [9]. The extracted juices are smooth, pulsating, and excellent in nutrients. These machines have a knob that you can adjust to achieve the necessary back pressure which gives more control over different ranges of produce with varying firmness as to making it more efficient at extracting a lot of juice. Triturating juicer is similar to single-gear juicers, with additional functions like it has separate kits for noodles making, nuts and seeds grinding, and chopping vegetables while in other juicers they are all incorporated as a single unit. Triturating juicers are mostly heavy and bulky, mainly desired for commercial activities.

2.4 Press Juicer (Citrus Juicer)

Citrus juicers are of several types: vertical hand-press type, pneumatic or hydraulic juice press type, press bowl type, and spinning bowl type, with others not mentioned. They are produced both as manual and electric juicer at different prices, categorized in every shapes, sizes, and materials. Citrus juicers are mainly preferred for orange juice processing, but a good number of juicers can also process lime, lemon, grapefruit, and even pomegranate. This citrus juicer requires the fruits to be cut into half across the middle and then place on the juicer. Manual citrus juicers some of them have handle for pressing of the fruit and squeeze out the juice, while others have a cone like cup for pressing the fruit until it extracts the juice. The juicer can be considered as cold-pressed processing in absence of heat, even though they are motorized machine [10]. These juicers most of them have detachable parts, which make them very easy to couple, dismantle, and tidy up. They are the more preferred juicer among families that make their own fresh orange juice because it takes them less than 3minutes to process and even clean up the mess from the juice. Omega, Black and Decker are the renowned manufacturers of this juicer.

CHAPTER THREE

3. MANUAL CITRUS JUICE EXTRACTOR OPERATIONAL ASSESSMENT

3.1 Hand presser/squeezer

The manual Juicers classified as hand juicing presser, are usually built with a rigid corrugated cone which presses down on half of any fruit be it orange, lime, lemon, tomato or other citrus fruit to extract its juice. This juicer requires your own muscle to squeeze the juices out from the fruits, but if it does not add heat to the juice, the result is a juice that is very nutrient-dense.

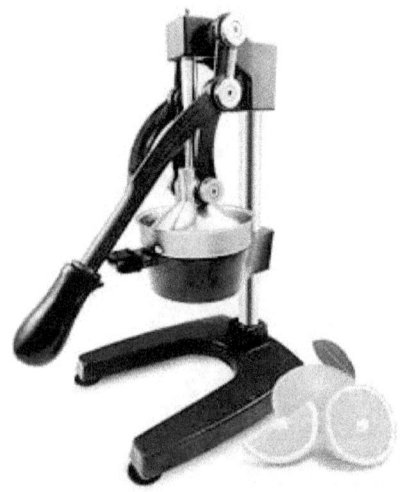

Figure 1: [a] Image of lever action Hand juicer

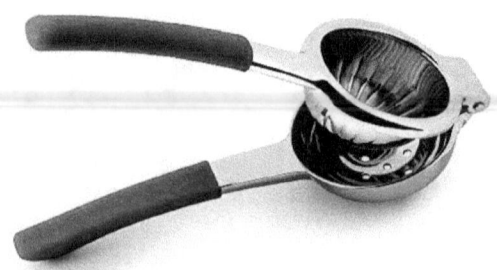

Figure 1 [b] Image of hand squeeze Juicer [8]

Hand Juice Press is a commercial grade juicer with a heavy-duty cast iron body and a 304 stainless steel strainer. It has a long, lightweight handle with a comfortable rubber grip to reduce any stress on your hand. This handle is in an upright position moving vertically downward at 90 degrees with a sturdy base that ensures no slipping or tilting during use. It has provided a safety hat to lock it into place so that it will not fall on you accidentally. This juicer is super-easy to use. It only takes three steps to process the fresh juice. First, cut your fruit into half, place the flat side down on the pressing plate, and press down the handle. The fruit juice is collected effortless. The strainer and funnel are both removable from the unit, making the units clean-up a breeze. This juicer leaves a lot of pulp in your juice. The overall dimensions are (220 x 180 x 370) mm and container diameter 120mm. See fig 3.1

3.2 Hand Squeeze Juicer:

The Stainless-Steel orange squeezer is a heavy, 378grams, stainless steel juicer with superior strength. It has the right size of cup for any small citrus fruit to fit into. Larger oranges and grapefruit are too big for this unit. We noticed that the cup showed some pitting after minimal use. This unit has long, thick, silicone-bonded handles to ensure the best comfort for your hand, and give you the best leverage to get the most quantity of juice. There is an issue with the handles, though. This orange squeezer is safe for easy cleaning, but the rubber on the handles is not molded on. That means if you throw it in the dishwasher, water will get in under the rubber coating and allow it to slip off. Juice squeezer has several models. See fig1b.

3.3 Hand- held Juicer:

It is a 71-gram unit with its own measuring cup, and a strainer built into the reamer. The reamer fits on the top of the measuring cup, but it does not lock in its position very well. You may have to help hold it in place with the same hand you are using to hold the cup. The measuring cup has a drip-free pour spout, so you can transfer your juice into whatever container you like without making a mess. The measuring cup and reamer is both Bisphenol A -free for your good health. The reamer can be removed from the cup for easy cleaning. The reamer is a bit easy to clean, but the cup takes a little more effort.

This juicer is pretty small, not leaving a whole lot of room for your hand to fit into. What makes it even more challenging is that the cup has a slightly narrowed neck, which makes the opening even smaller. It can still be named oxo good grip citrus juicer which is manufacturers' trade name.

Figure 2 [a] Image of Hand –held citrus juicer

Figure 2 [b] Component parts of dome shape hand juicer [11]

3.4 Dome shape juicer:

This juicer has multiple functions. If the reamer is facing up, it's suitable for juicing lemons, oranges, and other citrus fruits. When you flip the reamer over and have it pointing downward, it does excellently at juicing watermelon, pomegranate, and large fruits. The juice is caught in a cup that holds one cup of liquid. It is consisted of middle seat, Juice cup and upper cover. This juicer is easy to assemble and take apart to clean, but it does have a lot of pieces to keep. See Fig 2b

3.5 Cup Hand- held citrus Juicer.
It is not a standard juicer. The top cup helps to hold your fruit, while you can use the collection container for drinking of juice. The lid can be flipped over, so that it stores neatly inside the cup, but it can also act as an egg separator when in this inverted position. The unit consists of two different sizes of reamers, one 12-tooth reamer developed to extract juice optimally by fully pressing each lemon, lime or any other smaller citrus fruit; and the other is one 3-claw large reamer for juicing bigger fruits and orange. There is also a large 476gram capacity cup for collection of juice. At times are referred to as manufacturer of the product like sunhanny orange squeezer.

3.6 An orange juice Extractor
A designed and constructed orange juice extractor machine with diameter of 160 mm and a height of 350 mm have small blades sharpened that is coupled to a shaft which rotates with the bevel gear drive mechanism to actualize the fruit extraction [12]. The turning of the handle rotates the machine designed for high efficiency and ease of operation, which combine the extraction and beating often by macerating. The orange juice extractor encompasses of two main component parts a goblet and a physically operated mechanism. The physically operated mechanism contains a pair of bevel gear, two bearings and two shafts all in a casing. The following components were fastened to make up the drive mechanism, handle, Small sharpened blades, impeller shaft, bearing, dynamic seal and the goblet for leak proof. The performance test of the extractor machine showed that about 180-220 oranges were extracted per hour.

Figure 3 (a) Illustrating the Cup hand-held citrus juicer operation

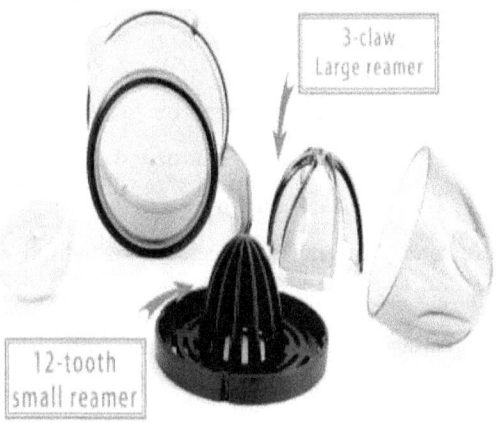

3-claw
Large reamer

12-tooth
small reamer

Figure3 (b) component parts of a cup hand-held citrus juicer [11]

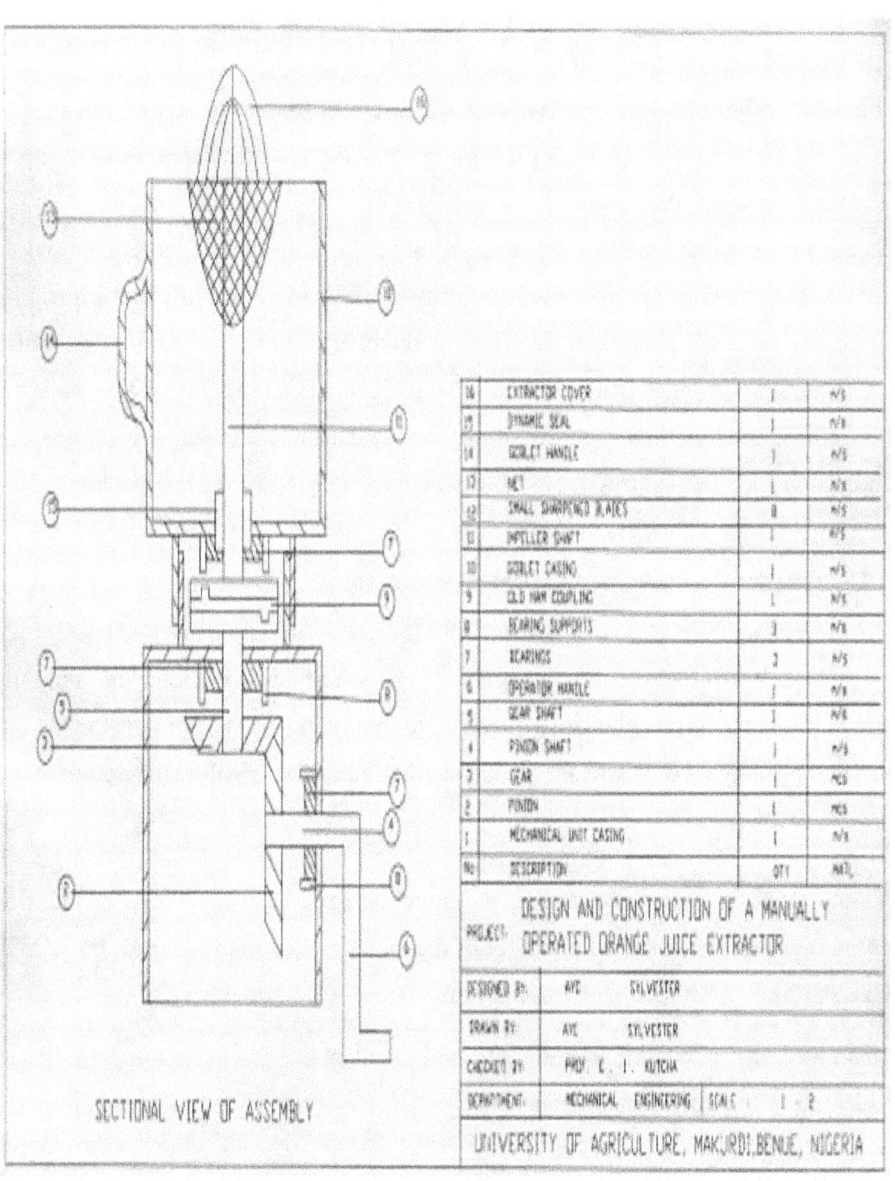

Figure 4: Manual orange juice extractor with the legend [12].

CHAPTER FOUR

4. DESIGN CONSIDERATIONS

For a good design of extractor machine, considerations to make includes: juice quality, processing material quality, high processing capacity and efficiency, availability and cost of construction materials.

Other considerations includes the need to make the core components with stainless steel because the juice are always in contact with them to ensure safety, corrosion free and quality of juice; to design the pressing chamber to hold the required quantity of raw materials (orange fruit) and the auger (worm shaft) to ensure maximum conveyance, crushing and pressing of raw materials. The main frame was considered to be strong for a better structural stability and strong support for the machine.

4.1 Criteria for designing a masticating extractor

4.1.1 Pulley and belt

For a V–belt to overcome slippage during power transmission, the maximum permissible ratio of diameter of shaft pulley to that of electric motor is 4:1

Therefore, the speed of worm shaft was determined from the following equation as:

$$\frac{N_1}{N_2} = \frac{D_2}{D_1} \qquad (1)$$

Where, N_1 is the rated speed of the motor in rpm, N_2 is the speed of the driven shaft (auger) in rpm, D_1 is the diameter of the motor pulley in mm and D_2 the diameter of the shaft pulley in mm. Belt Calculation

Equations 2 and 3 stated below will be used to calculate the centre-to-centre distance and length of the transmission belt:

$$C = \frac{D_1 - D_2}{2} - D_1 \qquad (2)$$

$$L = \frac{\pi}{2}(D_1 + D_2) + 2C + \frac{(D_1 - D_2)^2}{4C} \qquad (3)$$

where, C is the centre-to-centre distance and L is the length of the belt,

4.1.2 Torque

The torque was derived from the following expression as contained in machine design textbooks [13]:

$$T = \frac{2\pi N}{60} \qquad (4)$$

Where N is the speed of Conveyor (auger shaft) in revolutions per minute (rpm)

4.1.3 Auger (Worm) shaft

The auger shaft which conveys crushes and presses the fruits inside the extraction chamber. The diameter of the auger shaft was determined from the equation given in design textbooks [13] as:

$$d = \sqrt[3]{\left\{ \frac{16}{\pi \tau_{max}} \times T \right\}} \qquad (5)$$

Where, d is the shaft diameter, T is the Twisting moment (torque) acting upon the shaft, τ_{max} is maximum shear stress and π is a constant.

4.1.4 Machine capacity

The theoretical capacity of the machine was calculated by the equation given by as follows:

$$Q = 60 \times \frac{\pi}{4} (D^2 - d^2) p \, N \, \varphi \qquad (6)$$

Where, Q is the capacity of the auger (m³/h), D is the screw diameter (m), d is the shaft diameter (m), p is the screw pitch (m), N is the shaft (rotational) speed (rpm) and φ is the filling factor.

4.1.5 Power requirement

The power required to drive the machine was calculated using an equation adapted from as:

$$P = Q \times g \, (LK_I \pm H) \, K \qquad (7)$$

Where, P is the power required to drive the machine (Watt), Q is capacity of the auger (kg/s), L is the auger conveyor length (m), g is the acceleration due to gravity (m/s²) Ki is coefficient of friction for fruits, grains and chopped hags); K is overloading coefficient, H = perpendicular height; and F is the material falling factor. To accommodate for other losses and the power used in driving the pulley, the mechanism requires an electric motor which satisfies the expression below:

$$P_m = \frac{P}{\eta} \qquad (8)$$

Where, P_m is the power of electric motor (W) and η is the motor efficiency.

4.1.6 Volumetric Capacity

The volumetric capacity of the machine is given by as expressed in Equation 9.

$$Q_{vc} = \frac{Q_e}{\rho} \qquad (9)$$

Where: Q_{vc} is volumetric capacity, Q_e= the theoretical capacity of the extractor, ρ is the density of fruit (kg/m³).

4.1.7 Weight of Material (Orange Fruit)

The weight of the orange from which the juice are obtained is calculated using the expression,

$$q_m = \frac{Q}{V} \qquad (10)$$

Where, q_m = weight of the materials to be transported (kg/m), V is the velocity of the auger (m/s).

4.1.8 Velocity of the auger

$$V = S \times \pi \qquad (11)$$

n = number of screw rotation and is taken according to the conveyor materials, S = Pitch of the auger

14

4.2 Design criteria for Triturating extractor machines

4.2.1 Gear Design and Analysis

The material used for the manufacture of gears depends upon the strength and service conditions like wear, noise etc. Metallic and non-metallic materials could be used for gear manufacturing. Metallic gears with cut teeth are commercially available in cast iron, steel and bronze, while the nonmetallic materials like wood, rawhide, compressed paper and synthetic resins like nylon are used for gears, especially for reducing noise. Cast iron is extensively used for the manufacture of gears simply because of its good wearing properties, excellent machineability and ease of producing complicated shapes by casting method. Cast iron gears with cut teeth may be engaged, where smooth action is not essential. Steel is widely considered for high strength gears and may be plain carbon steel or alloy steel. The gears are usually heat treated in order to achieve combined property of toughness and hardness of the tooth.

4.2.2 Gear Design Data

In the design of a gear drive systems, the following data is usually given:
- The power to be transmitted.
- The speed of the driving gear,
- The speed of the driven gear or the velocity ratio, and
- The centre distance.

4.2.3 Gear Requirements

The following requirements must be met in the design of a gear drive system:

(a) The gear teeth should have sufficient strength so that they will not fail under static loading or dynamic loading during normal running conditions.

(b) The gear teeth should have wear characteristics so that their life is satisfactory.

(c) The use of space and material should be economical.

(d) The alignment of the gears and deflections of the shafts must be considered because of its effect on the performance of the gears.

(e) The lubrication of the gears must be satisfactory.

4.2.4 Design of Spur gear

In order to design spur gears, the following procedure may be followed; the tangential tooth load is obtained from the power transmitted and the pitch line velocity by using the following relation:

$$W_T = \frac{P}{v} \times C_s = \frac{60000 \times P}{\pi d N} \qquad (12)$$

Where

W_T = Permissible tangential tooth load in Newton

P = Power transmitted in watts

v = Pitch line Velocity in (m/s); $(v = \frac{\pi d N}{60})$

C_s = Service factor = 10^3

We know that circular pitch

$$P_c = \frac{\pi d}{T} = \pi m \qquad (13)$$

$d = mT$

Where

P_c = circular pitch

T = Number of teeth

m = Module in meters; $(m = \frac{d}{N})$

d = Pitch circle diameter in meters

Thus, the pitch line velocity may also be obtained by using the following relation;

$$v = \frac{\pi d N}{60} = \frac{\pi m T N}{60} = \frac{P_c T N}{60} \qquad (14)$$

Where

N = Speed in revolutions per minute

4.2.5 Maximum Tooth Load

It is the amount of load the teeth can carry without premature wear. The following expression as applicable in design textbooks:

$$W_W = D_p b Q K \qquad (15)$$

16

Where W_w is the Maximum or limiting load wear (N), D_P is Pitch circle diameter of the pinion (mm), b = Face width of the pinion (mm), K = Load-stress factor (N/mm²). Q is Ratio factor

Also, using torques of the pinion and gear;

$$W_W = \frac{2T_G}{T_G + T_P} \qquad (16)$$

V.R. = Velocity ratio = T_G / T_P,

4.2.6 Torque

Pinion torque is a convenient criterion for approximate rating of level years, requiring conversion from power to torques by the relation.

$$T_P = \frac{9550P}{N_P} \qquad (17)$$

Where

T_P = Pinion Torque (Nm)

P = Power, hp (KW);

N_P = Pinion speed, rpm.

4.2.7 Gear Ratio

The desired gear ratio is determined by the designer from the given input speed and expected output speed. It is also called velocity ratio and defined as the ratio of rotational speed of input gear to that of the output gear.

$$V_r = \frac{N_g}{N_P} = \frac{D_g}{D_P} \qquad (18)$$

Where

N_P = Number of teeth on pinion,

N_g = Number of teeth on gear

D_P = pitch diameter of pinion,

D_g = Pitch Diameter of gear

4.2.8 Compound Gear Train

A gear train is when two or more gears are working together by meshing their teeth and turning each other in a system to generate power and speed.

$$V_r = \frac{Gear\ A}{Pinion} = \frac{Gear\ B}{Gear\ A} = \frac{Gear\ C}{Gear\ B} = \frac{Gear\ D}{Gear\ C} = \cdots$$

Multiplying the ratios together we get the total gear ratio.

$$\frac{N_2}{N_1} \times \frac{N_5}{N_2} \times \frac{N_7}{N_5} = \frac{N_7}{N_1} = n \qquad (19)$$

Where

N_1 = Number of teeth in gear 1

N_2 = Number of teeth in gear 2

4.3 Electric power operated orange Juice Extractors

4.3.1 Masticating machines

The mass of juice in waste product was ascertained using the technique of American Society of Agricultural Engineers (1982), which involved oven drying the chaff at 130°C until a constant weight was reached. A stop watch and weighing balance were both used to obtain the time of extraction and measure the mass of the extracted fruit and chaff. The experiment was replicated five time using orange, while for multipurpose extraction machine it is replicated thrice for each fruit. The test was carried out at different extraction speeds with the aid of gear arrangement. The juice yield, extraction efficiency and extraction loss of these machines were calculated using equations (4.1 – 4.3) as stated below

4.3.1.1 Juice yield equation

$$J_Y = \left(\frac{M_{JE}}{M_{JE}+M_{RW}}\right) \times 100 \qquad (20)$$

4.3.1.2 Extraction efficiency equation

$$E_F = \left(\frac{M_{JE}}{XM_{FS}}\right) \times 100 \qquad (21)$$

4.3.1.3 Extraction loss equation

$$E_L = \left(\frac{M_{FS} - (M_{JE} + M_{RW})}{M_{FS}}\right) \times 100 \qquad (22)$$

Where:

M_{JE} = Mass of Juice extracted (kg)

M_{RW} = Mass of Residual waste/dry pulp (kg)

M_{FS} = Mass of Feed sample (kg)

J_Y = Juice yield (%)

E_F = Extraction efficiency (%)

E_L = Extraction loss (%)

The juice constant was calculated by getting the ratio of the sum of masses of juice extracted and juice in chaff to the mass of fruit feed sample.

4.3.1.4 Juice constant of Orange fruit

$$X = \left(\frac{M_{JE} + M_{JC}}{M_{FS}}\right) \qquad (23)$$

Where:

x = Juice constant of fruit (decimal)

M_{JC} = Mass of Juice in the chaff

4.3.2 Juice pulping machine

A juice pulping machine is an electrically powered juicer [14]. It consists of an auger-sieve combination placed above an aluminum frame, a handle for manual operation and produces juice free of seed and skin. The fruit press consists of a crusher mounted on components like screw-thread, crusher and slated cage. The machine is a lever operated press that grinds and crushes in one operation with an output of about 25 litres of juice per hour when operated by one person. The machine is using a masticating process.

Figure 5: Image of Juice pulping machine [14]

4.3.3 Juice Extractor machine

The juice extractor machine has a power requirement of 1.17 kW and is operated by a 1420 rpm electric motor [15]. Both the extraction capacity and extraction efficiency of the extractor machine were performed. With an average juice extraction capacity for orange as 5.10 kg/hr and its extraction efficiency as 78.78% while that of grape is 2.79 kg/hr and 75.66 % respectively. 280 % of the manual extraction approach was reported as orange juice extraction capacity while 304 % of the values obtained were activated by using a domestic extraction cup. The manual extraction method accounted for 220 % grape juice extraction, with 180 % of the value obtained necessitated by the use of the extraction cup. The straight auger introduced as a replacement for tapered one resulted in an increase in the juice extraction efficiency of 89.2% and juice extraction capacity of 15.8 kg/h for sweet orange. The shaft speeds of the machine were studied with sorted grades of fruits according to its size and thickness with respect to extraction efficiency and capacity using regression analysis. The various fruit thickness used for the test was 20, 40, and 60 mm with extractor shaft speeds of 300, 400, 500 and 600 rpm. Three categorical sizes of pineapple and orange fruits studied showed that juice extraction efficiency and capacity have very strong quadratic relationships with speed. An exceptional linear relationship exists between extractor shaft speed and its capacity for 60 mm apple thickness.

4.3.4 Juice Extractor machine

Juice Extractor machine has been developed with the performance evaluation conducted as a function of its extraction efficiency [16]. The extractor components parts are as follows: screw jack, screw connecting rod, pressing mechanism, frame, interlock, hopper, and discharge mechanism. The performance evaluation tests showed an improved juice yield of 76%, with extraction efficiency of 83% and low extraction loss of 3%. It is a masticating machine with the same basic component parts.

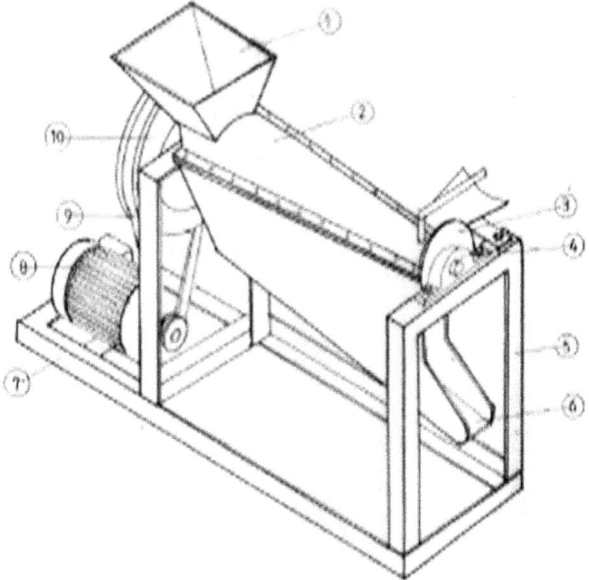

Figure 6: Mini orange juice extractor with component parts as: 1-Hopper; 2– Extraction compartment; 3– Disc plate; 4– Bearing housing; 5– Frame support; 6– Juice conveyor; 7– Base stand; 8– Electric motor; 9– Transmission belt; 10– Pulley [17]

4.3.5 Mini Orange juice extractor

Mini orange juice extractor was constructed with locally available fabrication materials for a small scale processing [17]. The mastication machine components parts includes hopper for feeding in orange, a lid cover, auger shaft, strainer, juice conveyor, trash outlet, transmission belt, frame support, pulleys and bearings. During processing, the auger shaft transmits, compresses, squashes and constrict the fruit to extort the juice which in turn passes through the juice sieve for filtration and then to the juice conveyor while the remaining waste is pushed to a trash bin. An average juice yield of 41.6 % and juice extraction efficiency of 57.4 % were recorded. The machine has a 2 horse power electric motor with juice extraction capacity of 14 kg/h. See figure 6.

4.3.6 A motorized Fruit Juice Extractor

A motorized fruit juice extractor machine was developed with the orange fruits washed and weighed (as 1kg, 1.5kg and 2kg respectively) of fruit sliced into 8 and 16 parts using the extractor to process the juice [18]. The juice yield, extraction loss and extraction efficiency of the machine were obtained using equations 4.1- 4.3 above. Normal juice yield of 64.6 % extraction efficiency of 68.2 % and corresponding extraction loss of 7.05 % were obtained from the 16 slice lengths of orange fruit. From the test result carried out using the juice extractor and the hand squeezing method, it was obvious that the rate of extraction increases as the weight of fruit increases with a corresponding increase in the juice yield and extraction efficiency. The juice extraction efficiency average and capacity were 57.70 % and 25.83 % respectively. This study reveals that juice yield and extraction efficiency reduces while extraction loss rises with amplified size of fruit slices. Juice yield, extraction efficiency and extraction loss from 16 slice lengths oranges ranged between 48.90 – 64.60 %, 50.00 – 68.20 % and 0.6 –7.35 % respectively. The higher extraction efficiency (mean value) of 57.70 % of the juice extractor reported the extraction rate to be more proficient than that of the hand squeezing method which has extraction efficiency (mean value) of 28.5 %. The motorized juice extractor components includes Hopper, Transmission Belt, Power Shaft coupling, Bearing, residual Outlet, Juice conveyor, Shaft housing, Seal, Cylindrical Drum, Electric Motor, Bolt, Adjustable Port and Frame Support.

4.3.7 A Multi-Fruit Juice Extractor

Multi-fruit juice extractor design and construction with performance evaluation on fruits of pineapple, orange and melon were conducted [16]. The extractor functions on the principle of compressive and shear squeezing force wielded through an auger transmission system. The associated component parts consists of a tool frame, collection channel, gear box, juice extraction chamber, tapered auger shaft, perforated screen base, and electric motor. The components design analysis provided the parameters incorporated in the sizing, fabrication and coupling of the machine. Performance evaluation of peeled or unpeeled fruits of pineapple, orange and water melon indicated percentage juice yield of 79.1, 68.7 %, and 77 or 69.2 %, 89.5 and 89.7 % respectively, while extraction efficiency of 96.9 %, 94.3%, and 96.6 % for peeled pineapple, oranges and water melon respectively and 83.6 %, 84.2 %, and 97.1 % respectively for unpeeled and extraction losses of peeled and unpeeled fruits of pineapple, oranges and water melon are 2.1 and 2.7 % , 2.1 and 2.5 %, and 2.9 and 2.6 % respectively. The machine is recommended for households and local fruit juice vendors because its operation is simple, easy and maintenance friendly. See figure 7.

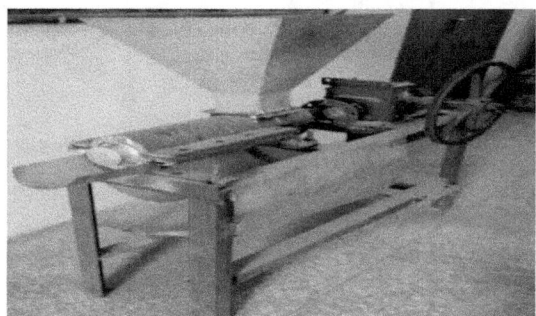

Figure 7: Image of a multi Fruit juice Extractor [16]

4.3.8 Mechanized Fruit Juice Extractor

This extractor machine serves as both slicer and extractor of fruits and vegetable [19] with the assistance of the slicing blade, screw conveyor shaft, hopper, electric motor, gear train, conical resistor, juice collector, waste collector, barrel and ball bearings it exerts contact shear and compressive force. The fruits fed into the machine are continuously crushed by a metal crusher against the metal surface that separates the juice from the waste which then is collected through a unique channel while the wastes are pushed to the trash bin. The efficiency of the machine is 67% output and the throughput of 4.8 litres per Minute. It is a masticating machine with a robust construction and convenient design. It appears as table top machine with the dimensions as 500mm x 300mm, therefore there are available for domestic and commercial activities.

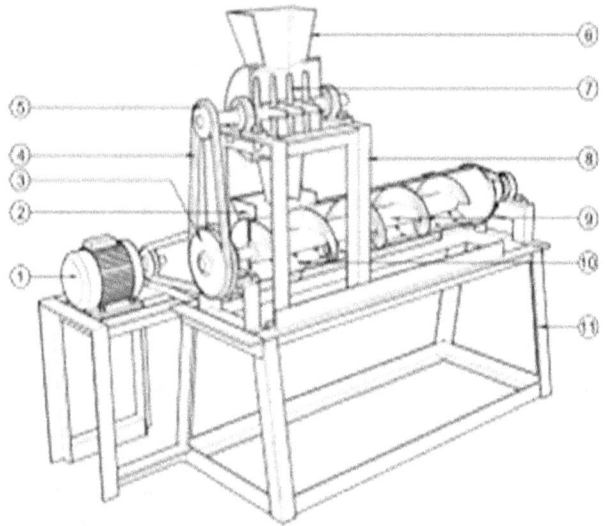

Figure 8: Image of Fruit juice extractor machine [20]

4.3.9 Fruit juice extractor machine

This machine is divided into two basic compartments: the chopping and the juice extracting compartments [20]. The performance evaluation results reported a typical juice yield of pineapple, orange and ginger were 74 %, 72 % and 34 % respectively; while juice extraction efficiencies of 84 %, 80 % and 71 % respectively; and juice extraction losses of 18 %, 16 % and 9 % respectively at optimum machine speed of 335 rpm for pineapple and oranges whereas 476 rpm is for ginger. This Extractor with 3hp electric motor has the capacity of process 30 litres/hr of oranges. See figure 8.

4.3.10 *Modified Fruit juice Machine*

A manual fruit juice machine fabricated by Onyene was modified by the addition of electric motor [21]. Then performance evaluation to ensure that the already existing extraction parts can work well with the new amount of power that the motor produces when compared with human power. The auger transmits, crushes, presses and constricts the fruits to extract the juice. The juice extract passes through the sieve for filtration and collection in the juice collector while the waste is trashed into the basket. When tested for freshly harvested orange and pineapple fruits, results show that the normal juice yield for orange and pineapple were respectively 23.20% and 24.75 % as against 17.47% and 17.50% of the manual extractor; juice extraction efficiencies were respectively 60.22% and 65.76% as against 50.32% and 53.76% of the manual extractor; and juice extraction losses were respectively 12.86% and 14.04% as against 12.06% and 11.34% of the manual extractor at 1.2kg/min feed rate. The modified machine was calculated to be 21.04% more efficient than the manual operated machine. This extractor is powered by a 3hp electric motor with a process capacity of 16.2 litres/hr of orange. See figure 9.

Figure 9: Image of Hand modified Fruit juice Extractor [21]

4.3.11 Motorized Juice Extractor machine

A motorized juice extractor machine was developed and evaluated [22] with the results of the evaluation of fruit yield for orange, pineapple and golden melon as 38.00, 51.43 and 39.67% respectively, with extraction efficiencies of 65.47, 83.94 and 53.17% respectively while the juice extraction loss of 6, 9 and 28% respectively for the three fruits mentioned above. This extractor component parts includes the hopper, slicing chamber, extracting chamber, frame support and channels for juice and waste discharge. The juice extraction machine is powered by 2 horse power electric motor. See figure 10.

Figure 10: Image of a motorized juice extractor [22]

4.4 Centrifuge juice extractor machine

4.4.1 Centrifugal Beach Juicer Machine

It is a Centrifugal machine with a 76mm Big Mouth Feed Chute, power capacity 800W. The units comprises of orange reamer, strainer, knob, bumper, cup and its support kits, gear rack, under counter pulp chute, and micro mesh metal strainer and blade set.

Figure 11: Images of a centrifugal juice extractor [23]

4.5 Triturating and Citrus orange extractor machine

4.5.1 Automatic orange juice extractor

An Automatic orange juice extractor is a 120 watt electric juicer which processes an astounding 22 to 30 oranges per minute. It weighs between 44kg - 46 kg as durable machine with the dimensions 508 x 890 x 508 mm, and is enclosed in a corrosion resistant, stainless steel housing. Although it is an 'orange' juicer, but can still be used for your other citrus fruits extraction, such as limes, grapefruit or succulent lemons for lemonade. This juicer has a strainer that collects the pulps and seeds as well as filter for your fresh juice, giving you the juice, and keeping the rest in a waste basket. This juicer has a very easy operational and cleaning approach with a safety cut-off switch. In operation, fruits are introduced into the machine through the hopper. They are received by the orange collector and sliced into two halves by means of a knife placed between the two collectors. Each half of the orange enters the collector is then pressed by the rotary balls. Thus, the machine transports, slices and presses the fruit inside the extraction chamber until juice is pressed out of the fruit. The juice extracted is drained through the perforation provided at the bottom of the extraction chamber. The halved squeezed orange residual waste is thrown out through the pulp outlet on both sides of the machine. The component parts of this machine included gear drive, Reamers (knaggy balls), Remaining collectors, knife, sieve, hopper, waste bucket and electric motor.

Figure 12: Image of an automatic orange juice extractor machine [24]

CHAPTER FIVE

5. CONCLUSION

The orange juice extraction machines have existed for a very long time now but the limitations associated with the once in use have raised concern for a review and best way to harness the fruits available during its season. Orange juice extraction is still a serious issue that desires attention, since a greater percentage of the fruit turns as waste during its season. In Nigeria for instant there is a large scale cultivation of orange fruits, which requires a suitable mechanism to be identified and developed for the fruit optimal harvest. Several attempts have been made in the past to mechanize the extraction of orange juice. Those attempts produced both manually and electrically operated machines. Some of the machines were gigantic, uneconomical, time and energy consuming. There is a great need to explore these existing mechanisms which apparently would be both economical and satisfactory to end users. Depending upon the needs of farmers and consumers, a suitable mechanism must to be selected.

REFERENCES

[1] Mushtaq M. (2018) "Fruit Juices Extraction, Composition, Quality and Analysis" Academic press, 131-159. https://doi.org/10.1016/B978-0-12-802230-6.00008-4.

[2] Farnworth E. R, Lagace M. , Couture R., Yaylayan V., Stewart B.. (2001) "Thermal processing,storage conditions, and the composition and physical properties of orange juice" .Food Res Int 34:25–30,. https://doi.org/10.1016/s0963-9969(00)00124-1

[3] Jiang X. (2014), "Design and Research on Multi-function Juice Extractor" Advance Journal of Food Science and Technology **6**(6) pg 774 - 779,. https://doi.org/10.19026/ajfst.6.109

[4] Boylston T. D. (2010) "Temperate Fruit Juice Flavors" Handbook of Fruit and Vegetable Flavors 451 - 462, https://doi.org/10.1002/9780470622834.ch24

[5] Eyeowa A., Adesina B., Diabana, Tanimola O. A. (2017). "Design, fabrication and testing of a manual juice extractor of small scale applications". Current Journal of Applied Scienceand Technology **22**(5): 1-7. https://DOI: 10.9734/CJAST/2017/33360

[6] www.foodterms.com. "juicer : Encyclopedia : Food Network". Retrieved 13 January 2017.

[7] Ashurst P. R., (ed.),(1999) "Production and Packaging of Non-Carbonated FruitJuices and Fruit Beverages" Springer Science and Business Media New York. https://doi.org/10.1007/978-1-4757-6296-9

[8] https://www.google.com/search?q=manual orange juice extractor

[9] https://healthykitchen101.com/types-of-juicers/

[10] https://www.goodnature.com/blog/types-of-juicers/

[11] https://www.amazon.com/Sunhanny-Squeezer-Anti-Slip-Rotation-Transparent

[12] Aye S. A., & Ashwe A., (2012) "Design and Construction of an Orange Juice Extractor". Proceedings of the World Congress on Engineering Volume III WCE 2012, July 4 - 6, . London, U.K. ISBN: 978-988-19252-2-0; ISSN: 2078-0958(Print); ISSN: 2078-0966 (Online).

[13] Khurmi R.S, Gupta J.K (2011). A textbook of Machine Design, Eurasia Publishing House (PVT.) LTD

[14] Emelike, N., & Ebere, C. (2015). Effect of packaging materials, storage conditions on the vitamin C and pH value of cashew-apple (Anacardium occidentale L.) juice. Journal of Food and Nutrition Sciences, 3(4), 160-165.https://doi.org/10.11648/j.jfns.20150304.14

[15] Adewumi B. A., (2005) "Development Of A Manual Fruit Juice Extractor". Nigerian Food Journal 22(1), . https://doi.org/10.4314/nifoj.v22i1.33585.

[16] Odewole M. M, Falua K. J, Adebisi S. O, Abdullahi K. O, (2018) "Development and performance evaluation of a multi fruit juice extractor". FUOYE Journal of Engineering and Technology 3(1). https://doi.org/10.46792/fuoyejet.v3i1.171

[17] Olaniyan A. M., (2010) "Development of a small scale orange juice extractor". Journal of Science and Technology, 47(1), 105-108. https://doi.org/10.1007/s13197-010-0002-8.

[18] Bamidele C. S., (2011) "Design, Fabrication And Evaluation Of A Motorized Fruit Juice Extractor" B.Eng Thesis, Department Of Agricultural And Environmental Engineering, University Of Agriculture, Makurdi

[19] Gbasouzor A. I., Okonkwo C. A.,(2014) "Improved Mechanized Fruit Juice Extracting Technology For Sustainable Economic Development In Nigeria"Proceedings of the World Congress on Engineering and Computer Science 2014 Vol II WCECS 2014, San Francisco, USA. ISBN:978-988-19253-7-4, ISSN: 2078-0958(Print); ISSN: 2078-0966 (Online)https://doi.org/10.1007/978-94-017-7236-5_32

[20] Boih N. A., .(2015) "Design, development and performance evaluation of a fruit Juiceextraction machine", Unpublished M. Eng. Mechanical Engineering Project Report, Department of Mechanical Engineering, Ahmadu Bello University, Zaria, Nigeria.

[21] Nwoke M. C., (2017) "Modification & Performance Evaluation of An Existing Fruit Juice Extraction Machine,. https://www.academia.edu/,

[22] Omoregie M. J., Francis-Akilaki T. I., Okojie T.O. (2018) "Design And Construction Of A Motorised Juice Extractor" Journal of Applied Sciences and Environmental Management volume 22(2) 207 https://doi.org/10.4314/jasem.v22i2.9

[23] https://juicerkings.com/different-types-of-juicers/

[24] https://rainmachine.en.made-in-china.com/product/China-Commercial-Electric-Orange-Juicer-Extractor-Automatic-Slow-Lemon-Squeezer.html

Publisher: Eliva Press SRL

Email: info@elivapress.com

Eliva Press is an independent publishing house established for the publication and dissemination of academic works all over the world. Company provides high quality and professional service for all of our authors.

Our Services:
Free of charge, open-minded, eco-friendly, innovational.

-Free standard publishing services (manuscript review, step-by-step book preparation, publication, distribution, and marketing).
-No financial risk. The author is not obliged to pay any hidden fees for publication.
-Editors. Dedicated editors will assist step by step through the projects.
-Money paid to the author for every book sold. Up to 50% royalties guaranteed.
-ISBN (International Standard Book Number). We assign a unique ISBN to every Eliva Press book.
-Digital archive storage. Books will be available online for a long time. We don't need to have a stock of our titles. No unsold copies. Eliva Press uses environment friendly print on demand technology that limits the needs of publishing business. We care about environment and share these principles with our customers.
-Cover design. Cover art is designed by a professional designer.
-Worldwide distribution. We continue expanding our distribution channels to make sure that all readers have access to our books.

www.elivapress.com

www.ingramcontent.com/pod-product-compliance
Lightning Source LLC
Chambersburg PA
CBHW051257170526
45165CB00004B/1754